Ranger Rick® Presents
Amazing Creatures of the Sea

Amazing Creatures of the Sea

National Wildlife Federation

Copyright ©1987 National Wildlife Federation.
All rights reserved. Reproduction of the
whole or any part of the contents without
written permission is prohibited.

Library of Congress CIP Data: page 95

Contents

Life in Another World

Whoever named our planet *Earth* made a big mistake. A more accurate name would be *Ocean*, for more than 70 percent of the globe is covered with water. If you stand on a beach and look out to sea, somehow you know that a different world lies out there.

A big difference between life on land and life in the ocean is that animals on land are surrounded by air. We live at the bottom of a sea of air. Even birds and flying insects that take to the sky settle to the ground from time to time.

The ocean, on the other hand, teems with life in some places from top to bottom. With an average depth of about 2½ miles, the ocean provides much more space for life than the land provides.

Animals in the oceans follow one of three life styles. Members of one group, which includes jellyfish and the tiny young of some large fish, drift with the currents. They are unable to swim or too weak to swim against the currents. Those in another group, which includes most adult fish as well as squid, octopus, and whales, are the strong swimmers. Members of the third group, which includes corals and barnacles, stay put. They settle to the bottom or attach themselves to a stone or to a plant or to another

creature and stay there for the rest of their lives.

Some ocean creatures also have shapes unlike those of any animals on land or in the air. These strange creatures are built a lot like wheels. Starfish, for example, look the same from the front, the sides, and the back. Only their top and bottom surfaces are different.

In some ways, life is easier in the ocean than on land. Sea creatures rarely have to cope with sudden changes in temperature, as many land animals do. But at least one problem that sea creatures face is unique to life in water: sinking. Most of the ocean's creatures will sink to the bottom unless they can swim or have some built-in "life preserver" to hold them up.

Overall, the sea is a world of creatures often far different from the animals we know on land. The sea is also a world about which we still have a lot to learn.

Explorers in the sea find creatures of all sizes. Left, a large grouper lets a diver get close. Above, a school of tiny fish opens up as a diver passes through.

Fascinating Fish

Colorfully striped "sweetlips" often travel in large schools like this in shallow tropical seas.

What It's Like To Be a Fish

To be a fish means to swim through water. Most fish have streamlined bodies that help make traveling easier, but not all fish swim in the same way. Some can wriggle their entire body. Some, like the filefish (above), have stiff bodies and bend only the tail and the tail fin. Fish that are even stiffer, like the sheepshead (right), move by rowing with the fins just behind their gills.

What is it like to be a fish in the ocean? A fish's world there is often dark and chilly, and it is always salty and wet. How do ocean fish live in a world that is so different from our own?

Fish need oxygen just as people do. People use their lungs to get oxygen right from the air, which is 21 percent oxygen. Air has enough oxygen for everyone. Fish use their gills to get oxygen that dissolves in the water where they live. If that water doesn't have enough oxygen, the fish suffocate and die.

Do ocean fish drink salt water? Yes, they do. The liquids in an ocean fish's body are less salty than the ocean. As the result of a process called *osmosis*, the fish constantly lose water to the saltier sea. They must drink to replace the water their bodies lose.

People get sick if they drink a lot of salt water. So how do fish drink it without becoming sick? They get rid of some salt through their digestive tracts. They force out the rest through special cells in their gills.

Now think about how fish move. Some fish are able to walk on the ocean floor or even glide through the air. But most fish simply swim, swinging their tail regions from side to side to push their way through the water. A few fish can swim as fast as sixty miles per hour. The muscles some fish use when swimming are more powerful than those in the arms of any human weight lifter.

Not all fish swim the same way. When parrotfish swim, their front fins usually do most of the work, rowing along like oars. And seahorses swim by waving the fins on their backs. Seahorses wiggle their fins so fast the fins look like spinning propellers.

Some fish swim alone, but others travel in groups called *schools.* A school has been defined as a group of fish that are mutually attracted to each other. The name comes from a Dutch word meaning "crowds" and has nothing to do with the schools where people sit in classes. Some of the largest schools or groups contain more fish than there are people in the United States *and* Canada.

Whether fish are swimming or remaining still, the water's support makes them almost weightless. Even a 150-pound person weighs only 7½ pounds in the ocean. The water's support reduces the amount of energy fish must exert to stay at a particular level. In turn, that leaves them more energy for other activities, such as feeding, hiding, reproducing, or migrating.

Many fish can adjust their bodies' volumes to get the most support at different levels. The fish do this with special gas-filled organs called *air* or *swim bladders.* The bladders work like balloons. When a fish needs more support, it inflates its air bladder by forcing in gases from its bloodstream. When the fish needs less support, it absorbs those gases back into its blood. Fish that don't have air bladders—sharks, for example—must swim all the time, even when sleeping, if they want to keep from sinking to the bottom.

Air bladders help fish in other ways, too. In some fish they act

Food galore can be found in the ocean, but lots of it is protected by hard shells. This harlequin tusk fish has a simple way to get around that protection. It uses its strong jaws and teeth to crush shellfish and other mollusks.

like eardrums, picking up sounds in the water. In other fish they act like "voice boxes." Searobins, for example, are among the noisiest fish in the North Atlantic. They make their loud sounds by vibrating their air bladders.

Fish also have special organs called *lateral lines,* rows of sensitive pits along their sides. These pits are so sensitive to water pressure that they can detect the waves made by approaching fish or other objects. Lateral lines help a fish know when other creatures come near or when it is about to hit something in the dark.

Some fish also have keen senses of smell and taste. Sharks, for example, can detect a single drop of blood in 25 gallons of water. Fish living in deep or cloudy waters where it is hard to see probably use smell more than sight to find food. Some fish also have taste buds all over their bodies.

Imagine being able to stick your finger into a banana cream pie to find out how good it is!

Have you ever picked up a fish and noticed how slimy it feels? Slime makes a fish's body slippery, and that lets the fish move more easily through the water. Slime also protects a fish from harmful parasites and disease organisms that could grow on its body.

Most fish are covered with scales. Scales are a living part of a fish's body and add growth rings, just as trees do. By checking a fish's scales, experts can often tell how old it is, how many times it has spawned—reproduced—and sometimes how much it has grown each year.

Of course, there are exceptions. Most catfish, for example, have no scales at all. Their bare skin is protected only by their slime. And some types of carp have scales although other types do not.

No matter what you learn about how some fish live, you can almost always find others that live differently. After all, there are more than 20,000 kinds of fish in the world today. The important fact to remember is, fish are as much at home in the water as we are in our own backyards.

The blenny (far right) and the prickleback (right) live along coastlines, where they often hide in holes in rocks. The weird "plants" on their heads are flaps of skin that help to camouflage them.

Big things often are scarier than little ones, as the porcupine fish above seems to know. When it is threatened by another fish, the porcupine fish swallows water and puffs up like a prickly balloon (right). The water goes into a special pouch under the fish's stomach.

Is *That* a Fish?

Both the seahorse (above) and its Australian relative the leafy seadragon (right) live in seaweed and other sea grass. When they find a tiny shrimp or worm, they suck it into their long, vacuum-hose mouths. The leafy seadragon's "leaves" are really flaps of skin that help the seadragon blend into the seaweed.

Have you ever seen a sea creature that looked like a horse, or a worm, or seaweed? If so, you might have asked yourself, "Is that a fish?" The question is a reasonable one. The term "fish" includes many creatures that look nothing alikc. They have little more in common than that they live in water.

Seahorses and their relatives don't look at all like fish, but they are. They breathe through gills and they have fins. In addition, they are covered with bony plates that are modified scales.

Seahorses range in size from smaller than your little finger to larger than your foot. They swim upright and grab seaweed and grasses with their tails to hold themselves in one place while waiting for tiny shrimp and fish to drift by. They suck up these creatures into their tubelike mouths.

One of the most remarkable things about seahorses is the way they take care of their eggs. At spawning time, a female inserts a tube into a special brood pouch on the abdomen of a male and squirts her eggs into the pouch. Several females may deposit as many as 200 eggs into one male's pouch. About four to six weeks later, the newborn seahorses leave the pouch. They look just like their parents, but are smaller than the fingernail on your littlest finger. The tiny seahorses are on their own immediately after leaving the pouch.

Some of the seahorses' largest relatives are the seadragons. Found only off the coast of Australia, seadragons grow as long as this book is wide when you open it up.

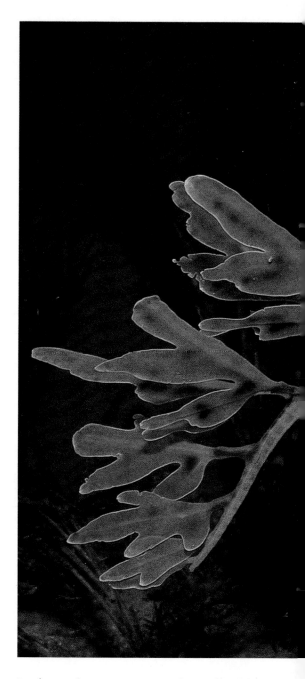

At first glance, you might mistake a leafy seadragon for seaweed. On much of its body it has flaps that look like leaves. These flaps help the fish blend in with the seaweed where it usually lives.

Skates and rays are another group of fish that don't look like

14

fish. All skates and rays are relatives of sharks. Their flat, wide bodies may be shaped like eagles and bats, guitars and saws, or disks and diamonds. Their teeth are short and flat for crushing shells, and some skates have flaps of skin that protect their eyes from bright sunlight when they are in shallow water—built-in sunglasses!

Skates are the largest group of these fish with flat, winglike fins. The smallest species is the Little skate. It weighs less than one pound and grows to be about 20 inches long. Along the Pacific coast

About 3 inches long and 1 inch wide, a *mermaid's purse* (right) is the egg case of a skate. A skate embryo (above) develops inside the case. The size of a dot at first, this embryo grew for about 63 days before the new skate hatched. The cases that people find on beaches are often empty.

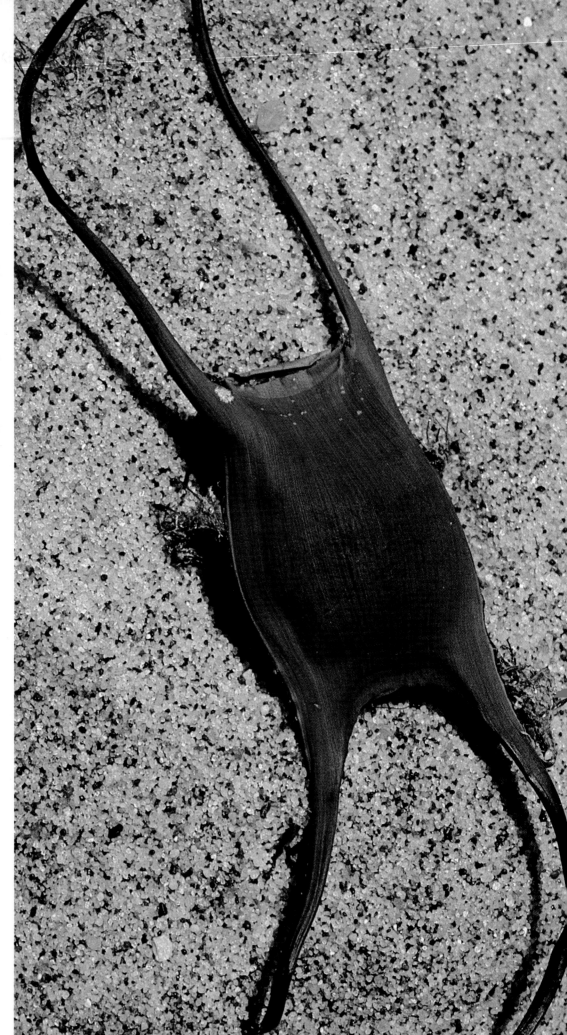

of North America lives the Big skate. These skates can grow to be more than eight feet long and can weigh up to 200 pounds.

Skates lay leathery, rectangular egg cases with prongs at each corner. The prongs anchor the egg cases to seaweed or rocks. After two to five months, the baby skates hatch and swim away. Storms wash some of the egg cases, called "mermaids' purses," onto shore where they dry up and turn stiff.

The manta ray is the giant of the skates and rays. The largest mantas are 20 feet wide and weigh almost two tons. A manta ray has two "horns" on its head, wide bat-like "wings," and a long tail. You can guess why this creature got the nickname "devilfish."

The manta rays' large size is really their only protection. Their small teeth are hardly used, and their "horns" are only soft flaps or fins that push small animals into their mouth. Some mantas have spines on their tails, but they usually are not poisonous.

It is the leaping ability of these gentle giants that may surprise you. Mantas start swimming very fast, burst into the air in spectacular leaps, and then hit the water with a loud *Boom!* Scientists aren't sure why mantas do this. There have been reports of females giving birth to their young during such leaps. But it is more likely that mantas leap to knock off pests, escape predators, or attract mates.

Another leaping ray is the eagle ray, named that because it looks like an eagle flying through the water. Like mantas, eagle rays can grow very large. One Florida

fisherman reported that a 300-pound eagle ray landed in his boat!

Eagle rays may swim like birds, but they feed like pigs. They push their pug-nosed snouts along the ocean floor to root out oysters, clams, or crabs. When they find something to eat, they squirt water at it to remove the sand. Then they crush the shells with their plate-like teeth, spit out the shells, and swallow the soft flesh.

Of all the rays, stingrays are the most dangerous. More people are seriously hurt by stingrays than by any other creature of the sea, even sharks. Some stingrays are as

Skates often live in shallow water where the sun shines down brightly. When they swim, they steer with their long tails.

A giant manta ray looks like it is flying when it flaps its broad, winglike fins. This fish feeds by swimming with its mouth open and swallowing small fish and other creatures.

small as a dinner plate, and others are as big as a bathtub.

Stingrays are named for the stinging spines on their tails, which they use only for defense. A stingray is harmless unless bothered or stepped on. When it feels threatened, it strikes with its long tail.

During the daytime when humans are about, stingrays are often hiding on the seabed, partly covered by sand. An unlucky person who steps on a stingray may be struck by a venomous spine on the stingray's tail. The wound is painful, but only rarely does the venom cause death.

A really shocking fish is the electric ray, which uses electricity to capture food. When a ray spots a crab or fish it wants to eat, it covers it with its winglike fins. Then it uses electric organs on either side of its head to send a deadly jolt of electricity through the prey. The largest electric rays can deliver shocks of up to 200 volts, almost twice the voltage of the electric current in your home.

Sawfishes are also a type of ray. Of all the rays, sawfishes most closely resemble sharks because of their overall shape. But sawfishes have something most sharks do not: a large "saw" on the front of their head. They use the saw as a combination rake and hoe. Sometimes sawfishes use it to dig in the sand or soft mud on the ocean floor to stir up crustaceans. Other times they swim into a school of fish and slash their double-edged saw from side to side. Then they return to eat the dead and stunned fish.

An eel may look like a slimy worm, but it is really a long, bony

Among the almost 600 species of eels are morays (right) and garden eels (far right). Morays live near the ocean bottom in reefs, and garden eels live on the sandy ocean floor in shallow water. The diver near the garden eels is entering a blind where she will hide to study these shy creatures.

fish. Small and delicate, garden eels live together in colonies on the ocean floor in a very unusual way. Each eel burrows tail-first into the sand and lives in the burrow. When hungry, the garden eels come out of their burrows and, standing upright on the tips of their tails, swoop gracefully to catch fish eggs and tiny animals that float by. When disturbed, they disappear almost instantly into their burrows.

Moray eels usually live along rocky coasts or coral banks. Some moray eels are fierce looking, especially the large ones that can grow up to ten feet long and weigh 75 pounds. They have powerful jaws and long, needle-sharp teeth.

The moray's thick skin is covered with mucus instead of scales. The mucus protects it from scrapes as it moves about searching for prey in cracks and crevices of the coral reef. Most types of morays hunt fish, but a few types eat crustaceans, mussels, and snails. In some of the morays that eat fish, the teeth point backward and keep wriggling prey from getting away when they start to eat it.

The bright colors of these fish help the fish blend in with colorful coral. They also help fish of the same species recognize each other.

A Living Kaleidoscope

One of the prettiest sights in the sea is a tropical reef alive with fish of almost every color and pattern. A reef is a ridge of millions of dead corals. The top of the ridge may be at the surface of the water or just below it. Such amazing animals as bright green triggerfish, blue tangs, yellow butterflyfish, black-and-white zebrafish, and schools of striped fish with the funny name of *sweetlips* (shown on pages 8-9) make their homes on tropical coral reefs.

Why are the fish that live around coral reefs more colorful than fish living elsewhere? That is an age-old question. Scientists have suggested several ways that flashy colors and bright patterns might help reef fish.

Sometimes a fish's bright color matches the color of the reef, and the fish blends right into the background. That helps the fish hide to ambush prey or avoid enemies.

Some fish have colors that stand out and serve as signals. One of the signals may be a warning that other fish should stay out of a fish's territory. Another signal may be a display to help fish of the same kind find each other.

In a few cases, however, the colors serve as advertisements that attract fish of different kinds. The neon gobie's bright electric-blue stripes, for example, attract large fish that let the smaller gobies pick off parasites or clean wounds. The large fish get cleaned up and the gobies get free meals.

Stripes can be useful in helping save a fish's life. Scientists have described what happens when a predator attacks a school of fish

which have horizontal stripes. The jumble of all the patterns together makes it harder for the predator to single out one fish to catch.

The reef offers many nooks and crannies where fish can go to hide. In fact, a photographer once spotted a parrotfish peeking out from a tiny cave which it had dug out of the coral.

Parrotfish don't usually dig their own caves, but they can do a lot of damage to a coral reef. Some scrape off coral particles to eat. Others may break off entire coral branches, chew them up, and eliminate what they can't digest. Some of this discarded, chewed-up coral builds up into deep layers of white sand around the reef.

Colorful reefs are very popular with people. One reef off the coast of Florida's Key Largo has been named an underwater state park. Another reef, near the Caribbean island of Saint John, forms the underwater part of Virgin Islands National Park.

The patterns of color on fish help disguise the shape of the fish and the position of the eye, making it hard for predators to spot them.

Hiding in Plain Sight

Did you ever wish that you could play hide-and-seek so well no one could ever find you? Some fish can hide right in plain sight. They have camouflage that helps them blend into the background. But hiding is not a game for fish and other animals. They do it either to get food or to avoid being eaten.

If you look at fish, you will see that many of them are light-colored on the underside and dark on the top. This pattern is called *countershading*. When a predator looks *down* through the water, it may not see its prey because the prey's dark top blends in with the darkness of the deep ocean. And when a predator is *below* its prey looking up, the prey's light underside blends in with the light coming from the surface of the ocean. The prey does not disappear completely, but it does become much harder to see.

You might think that fish with bright colors and bold patterns would be very easy to see. Many times they are. But these fish may also seem to disappear when they float motionless next to a colorful reef (see pages 22-25).

A single fish may also have different disguises at different times. One scientist has pointed out that the steelhead, a kind of trout, has countershading when it lives in the ocean. But when it leaves the ocean to live in shallow streams, it develops many new colors. These colors make patterns that seem to help the fish hide against the bottom of streams.

Flounders, flat fish that live on the ocean bottom, can change color to match different backgrounds. On a white bottom, flounders become light. On a black bottom, they turn dark. Scientists in a laboratory put a flounder on a checkered background to see what would happen. The fish developed light and dark patterns to match.

Though flounders, steelhead, and many other fish have colors and patterns that help them hide, other fish rely on their odd shapes. Leafy seadragons look more like seaweed than fish (see page 15), and long, thin pipefish look like leaves of eelgrass.

Camouflaged fish are often hard to find, but if you look closely at the picture of waving sea grasses (top), you will spot some thin bay pipefish. In the picture above, a flounder blends into the sandy sea floor. And that's not a lump of stone at right. It's an anglerfish.

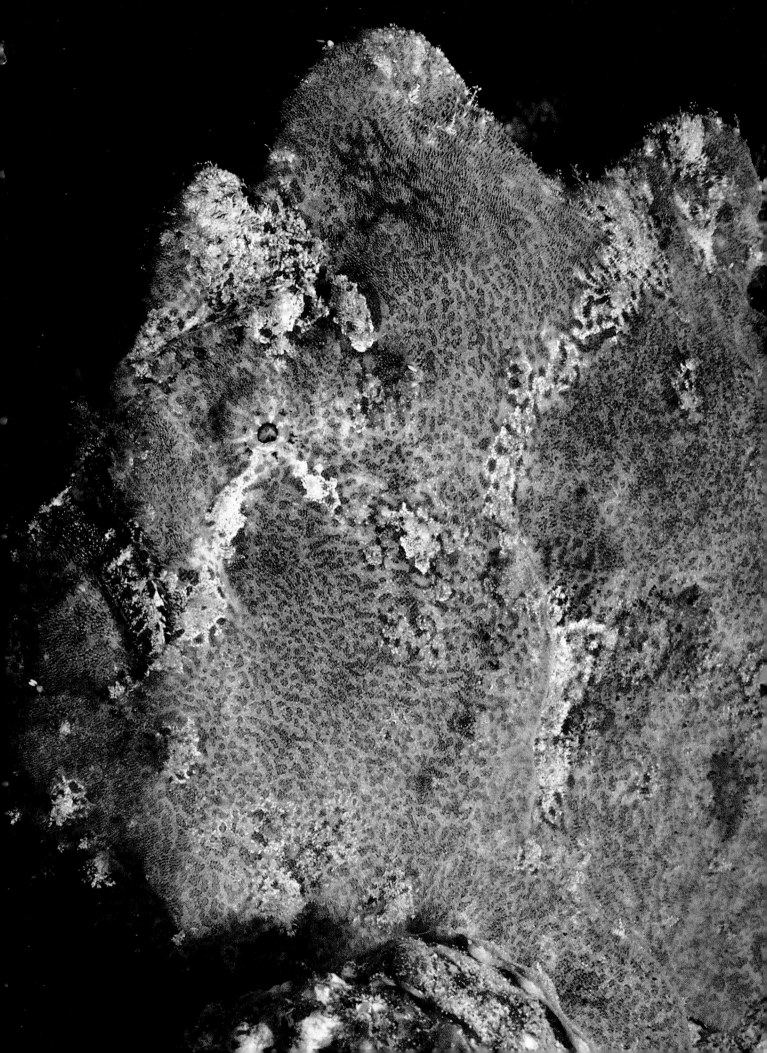

Fish That Walk and Fly

A batfish lurks on the ocean floor, walking with muscular fins. The bump in the middle of its head can extend into a long pole with a wormlike lure on the end. It uses this lure to fish for its dinner.

People are not surprised to see birds flying and dogs walking and running. That is what birds and dogs are "supposed" to do. Fish, on the other hand, are "supposed" to swim—and most of them do. But did you know that some fish can fly and walk, too? Here is how they do it.

The best-known flyer is the creature known as a *flying fish*. It has fins that look like wings. It cannot flap these fins the way a bird flaps its wings to fly, but it can stretch them out and glide.

When it gets ready to glide, a flying fish races through the water to build up speed. When it breaks through the surface, it may be going more than 30 miles per hour. A few extra flips of its strong tail fin in the water give the fish an extra push and it rises about three feet into the air. If it runs into a strong wind, the fish may rise several times as high.

Like any glider, a flying fish soon starts to settle to the surface. But all it takes is another push with its tail fin and it is off again. By gliding and pushing off several times, a flying fish may cover the length of two football fields before it settles back into the water.

Building up speed and breaking free of the surface is hard work. A

flying fish needs a heart large and strong enough to keep its muscles well supplied with oxygen-filled blood. Comparing heart size to body size, a flying fish has a heart twice as large as that of a mackerel or tuna. And mackerels and tunas are strong swimmers. Some cruise the seas at better than 30 miles per hour—twice as fast as a runner breaking the four-minute mile.

If flying—or gliding—takes so much energy, why do the flying fish do it? The answer is easy: to get away from danger. Tuna, which like to eat flying fish, can swim faster than their small prey. But tuna cannot fly. Flying fish that can reach the surface and glide usually get away. Sometimes, though, the flying fish still ends up in trouble. Sailors on small craft have caught and cooked flying fish that landed on their boats during the night.

Just as some fish use their fins to glide over the sea, other fish use their fins to move across the sea floor. Some small relatives of the whale shark live on the bottom, and when they are disturbed they walk or crawl away instead of swim. Lizardfish sit on the bottom, too, propped up on their front fins like people ready to do pushups. But these fish do not walk. Instead, they rest and wait for a likely meal to come by.

Like the lizardfish, batfish sit and wait for prey, too. And according to some reports, when batfish move they waddle across the bottom on both front fins and back fins—the batfish version of arms and legs.

Lizardfish may also use their fins to dig into the sandy bottom

A flying fish swims near Hawaii. These fish don't really fly. Traveling at about 30 miles per hour, they spread their fins and glide through the air for short distances.

when they want to hide. And sea-robins use parts of their long front fins as feelers, poking into the sea floor to find buried prey.

People once thought that the fish called *flying gurnards* had the best of both worlds, walking *and* flying. The gurnards' front fins are enlarged like the fins of the flying fish. And travelers reported seeing the gurnards gliding above the ocean's surface. Scientists now think that the observers really saw the familiar flying fish, not gurnards. Gurnards do use their strong front fins for crawling across the sea floor. And when spread out, the large fins may scare away the gurnards' enemies. But the name *flying*, however wrong, has stuck.

When a 2½-foot-long Red Sea grouper has eaten its fill of small fish, it seeks out a cleaner wrasse and opens wide. The hungry wrasse swims safely in and out of the predator's mouth, cleaning the larger fish's teeth and gills by picking out the leftover food.

Dynamic Duos

Sea creatures don't always live by the rule of "eat or be eaten." Some fish and other animals in the ocean live together in harmony. Sometimes they even help each other.

Cleaner wrasses (RASS-es) —bright, three-inch-long fish that live around reefs—work like your neighborhood cleaners. Their bright colors act as signs, attracting customers. Larger fish gather around to let the little fish clean their gills and scales. The area where the wrasses work is called a "cleaning station." By picking off parasites and old scales, the wrasses get something to eat. The other fish end up cleaner and healthier.

The wrasses may also earn their own safety. The large fish, which sometimes feed on fish the size of the wrasses, seem to leave the little cleaners alone. In fact, once scientist has called the cleaning station "neutral ground." He reports that fish that normally fight each other gather there in peace as they wait to be cleaned.

Sometimes that peace is broken by a little fish called a *sabertooth blenny*. The blenny looks and acts like a wrasse getting ready to clean another fish. But when the larger fish drops its guard and lets the blenny get near, it's in for a nasty surprise. The hungry blenny takes a bite out of its tail!

If some wrasses can be compared to *cleaners*, then remoras (REM-or-ahs) can be compared to *hitchhikers*. The flat top of a remora's head holds a suction disk that lets the fish stick itself onto sharks, whales, turtles, and even small boats. The remora gets a free ride when the other animal travels and may also get scraps of food left over from the other animal's meal.

Why do other animals let the remoras hang on? Scientists aren't sure. People have seen some remoras swim into the mouths of sharks and manta rays. Maybe the remoras work like wrasses, picking off pesky parasites.

Some fish find that teaming up with other animals gives them a safe place to hide from enemies. The small pearlfish that live in warm, tropical seas are examples. These unusual fish actually make their homes inside clams, oysters, and fat, worm-shaped animals called *sea cucumbers*. One kind of pearlfish is thought to spend its entire adult life inside a sea cucumber. But other pearlfish probably enter their hosts only when they need to get away from danger. They stay out as they search for mates or food.

The fish attached upside down atop the bigger nurse shark (above) is a remora. It is a hitchhiker, and it gets free rides by using the suction disk on its head to attach itself to sharks, large fish, and sometimes turtles and whales.

The clownfish and the poisonous anemone are partners. Other fish avoid the anemone's stinging tentacles, but the clownfish makes its home among them. It sleeps there, hides there to escape danger, and sometimes takes food out of the anemone's mouth.

Some pearlfish find their food right at home: They eat the insides of the sea cucumbers. This doesn't harm the sea cucumbers, though. Their insides grow back. And what do the sea cucumbers get in return? Nothing, as far as we know. Only the pearlfish seems to get any benefit from the relationship.

Sometimes the best place to find safety is right in plain view —but in a spot where enemies can't reach you. That is what clownfish do when they swim among the poisonous tentacles of sea anemones. If other fish get too close, the tentacles sting and paralyze them—turning them into dinner for both the anemone and the clownfish. At the same time, some people say that the clownfish scares away fish that would eat the anemone.

How does the clownfish keep from being stung? Sick clownfish *do* get stung. But healthy ones wear a thick coating of slime which protects them. Something in the slime may stop the anemone from using its stingers. Or the slime may lack a substance present in other fish that sets off the stinging cells.

Other fish find safety with poisonous jellyfish (see pages 64-65). The best-known are the *man-of-war fish* that swim around the Portuguese man-of-war. One lion's mane jellyfish may also house 200 or so tiny fish among its deadly tentacles. These "guest" fish get protection and sometimes food when they nibble on other fish killed by the jellyfish. In turn, the jellyfish may eat other fish that are attracted by their little guests.

Are the guest fish immune to jellyfish poison? Not completely, though some can stand ten times as much poison as other fish can. Mostly the tiny fish stay out of the way of the tentacles. And like the clownfish, some also have thick slime that protects them.

The three small fish above are right at home, protected from predators by the stinging tentacles of this jellyfish. Some scientists think that these small fish attract predators, which are then stung by the tentacles and eaten by the jellyfish.

33

Prize Catches

Black marlins live in the Indian and Pacific oceans. These feisty fighters can weigh 1,000 pounds or more. The record catch, taken off the coast of Peru, weighed over 1,500 pounds and was over 14 feet long.

For size, speed, and fighting action, few ocean fish are more prized by anglers than marlins, sailfish, spearfish, and their close relatives, swordfish. These fish all have long, swordlike bills and are called *billfish.*

A hooked billfish may leap wildly out of the water, dash from side to side, dive, and jump again. It may keep this up for hours before an exhausted angler can land this prize catch. It's easy to understand how tiring this struggle could be. The record-setting black marlin weighed more than 1,500 pounds.

Catching up with a billfish at sea is not easy. Some marlins regularly cruise at better than 40 miles per hour. And a sailfish was timed going more than 68 miles per hour. That's about as fast as a cheetah, the fastest of all land animals.

A 12-foot billfish's bill may be four feet long, a third of the animal's total length. Scientists have different ideas about what these bills are used for. The long "sword" may make the fish more stream-lined, helping it cut through the water. Or, when a hungry billfish attacks schools of small fish, its bill may serve as a weapon and stun or kill the prey.

People have found bills broken off in whales. And many reports tell of billfish running into boats. Were the fish really trying to use their bills as spears? Some scientists think these "attacks" were accidental. The fish may have been going so fast they couldn't turn in time when the whales and the boats got in their way.

Like many fish, billfish go through remarkable changes as they grow up. A female marlin may lay 30 million eggs, yet only a few are fertilized and survive to develop into adult fish. The newly hatched young are less than a half-inch long and look more like "bug-eyed monsters" than fish. They have oversized eyes, mouths full of teeth, and *no* bills.

Many of the billfish that become adults play important roles in people's lives. Sports fishermen try to catch them for trophies. And commercial fishing boats haul them in to provide food.

Some fishermen tag a billfish after it is caught and release it back into the sea. By noting where a tagged fish is released and later caught, scientists learn more about how far billfish travel. Some of the white marlin that spend the winter along the coast of South America move close to the Virginia shore for the summer. And some striped marlin tagged near Mexico have turned up southwest of Hawaii.

These fishermen off the coast of Australia are trying to land a black marlin (left). Striped marlin, blue marlin, and white marlin are also prize catches. Sailfish (above) aren't as big as marlins. The largest ones weigh a bit more than 200 pounds. But they have the most impressive dorsal (top) fin of all the billfish. The fin stays folded in a groove along the back except when the sailfish surfaces.

Deepest of the Deep

In the deep sea, lures—lighted (opposite) and unlighted (below)—and lights on the body (bottom inset) help fish catch food or find mates. The black swallower (top inset) can eat food bigger than it is, which helps when food is scarce.

Imagine a place where sunlight never brightens the total darkness, a place where the temperature almost never rises above 40°F. Such a place exists, and it covers half of our planet. It is the abyss, the deep sea, that part of the ocean below about 12,000 feet.

Because of the lack of light, no plants grow in the abyss. That means there isn't very much food available, which means fewer creatures can live there than in shallower, food-filled waters.

The creatures that do live in the abyss can't find a lot to eat, so the way they are built and the way they live must allow them to save energy. They are small—the largest is about a foot long. They grow and move slowly. And they have fewer young than their relatives closer to the surface. They also live longer. The bivalves—clams, oysters, and mussels—that you might find normally live for about ten years. Bivalves in the abyss of the North Atlantic may live for 250 years!

In the abyss, the only light comes from fish that glow. These fish are *bioluminescent.*

Some deep sea fish make their own light. Like fireflies, they create chemicals that glow when they mix. Other fish can't make light themselves. These fish have special pockets in their bodies where they carry bioluminescent bacteria. The fish look as if *they* are glowing, but it's really the bacteria.

Some glowing fish use their light to catch prey. They dangle their lighted lures like bait on a fishing line. Curious creatures that come too close end up as the glowing fish's dinner. Glowing fish may also use their lights to find each other for mating.

Not all deep sea fish glow. But all of them have special adaptations for living so deep in the sea. They have developed extra-good senses of sight and smell, which help them find each other. And some of them have become hermaphrodites (her-MAFF-ro-dites), which means that each creature is both a male and a female. That way it doesn't matter if one is a male and one is a female when two of them meet to mate.

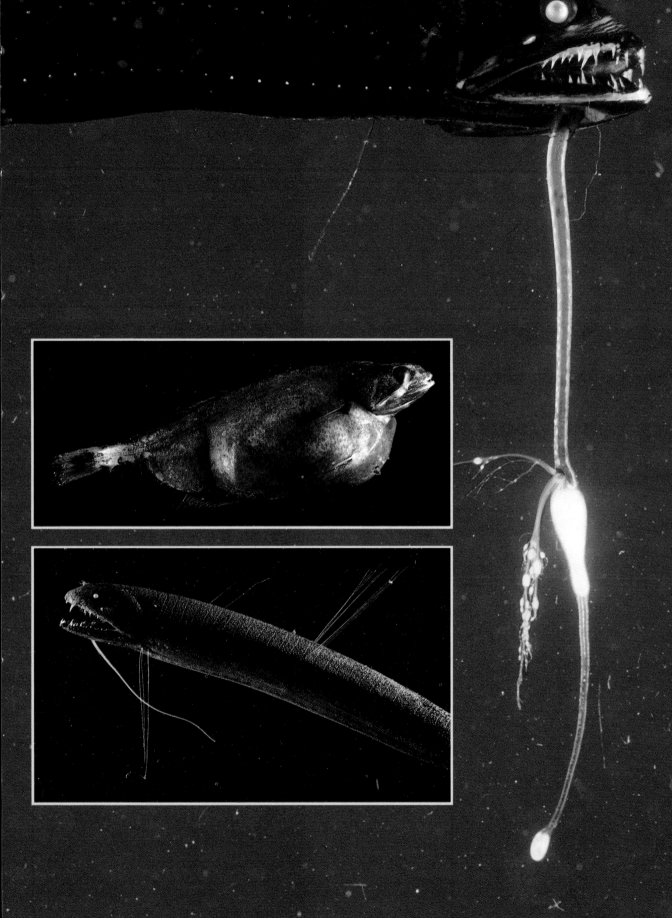

Look, but don't touch. The beautiful lionfish can inject poison into wounds caused by its spines. The lionfish also uses the fins on its sides as nets to help it catch prey.

Look Out! Danger!

Caution is a good watchword for divers exploring reefs. Many of the unusual fish to be found there are colorful, and most of them are harmless. But in some places, divers might come across creatures like the lionfish at left, an animal that is pretty but highly venomous. Or they might spot a silvery barracuda known as the great barracuda. That sharp-toothed predator may attack and bite without warning.

The lionfish belongs to a poisonous group of fish called *scorpionfish*. These fish have sharp spines that can give other fish, or a person, a painful wound. But to make matters worse, these fish also have special glands that sometimes inject poison into the wounds.

Lionfish swim around searching for prey among the reefs of the Indian and western Pacific oceans. When a lionfish sees small fish it wants to eat, it spreads its side fins like a net and drives the fish into a corner in the reef where it can catch them. The lionfish may also attack prey, and divers, with its venomous dorsal (top) fins.

The lionfish itself is probably safe from many predators, though. Scientists say that its bright patterns could be a signal that says "I am dangerous. Stay away!"

The scorpionfish, which has the same name as an entire group of fish, is almost as venomous as the lionfish. But this creature of the Mediterranean Sea and the eastern Atlantic Ocean uses its venom only to protect itself. Otherwise, this quiet fish sits alone, waiting for medium-sized fish and other prey to wander by. When one does, the

scorpionfish may dart out, snag its prey, and settle down for a meal. This fish has many relatives along both coasts of North America.

Stonefish are probably the most dangerous of the scorpionfish group. Like the lionfish, they live in the Indian Ocean and the western part of the Pacific. They have the most powerful venom, and they are among the hardest to see. When these two-foot-long creatures lie motionless on the rocks, they look just like stones. That's how these fish got their name.

A careless wader who steps on a camouflaged stonefish may be in real trouble. If the stonefish raises its stiff back fins, the tips of the fins may go through a rubber-soled shoe and right into the wader's foot. The poison that enters the wound causes extreme pain and sometimes death.

Just as stonefish present hazards to waders, great barracudas can be a problem for divers. Great barracudas can grow to be six feet long. They have extremely sharp teeth. A great barracuda's front teeth are longer than the other ones and are used to hold onto prey it attacks. If the prey is too large to swallow whole, the barracuda slices it into pieces with its other teeth.

The problem for divers is the great barracuda's curiosity. These fish may follow swimmers, boats, and even people walking on the shore. Scientists say the fish are attracted by shiny objects and by motion and aren't really after people. Nevertheless, great barracudas have been known to attack divers. Attacks by other kinds of barracudas are extremely rare.

Looks can deceive! The toothy barracuda (left, top) looks fierce, but barracudas rarely harm humans. Yet the camouflaged stonefish (left) and scorpionfish (top) —which look weird, not dangerous—are poisonous and can seriously harm anyone who steps on them.

Shark!

This sand tiger shark's teeth fall out when they become worn or snagged on prey. So do the teeth of all other sharks. Lost teeth are quickly replaced by new ones lined up right behind the old ones.

Few creatures of the sea are as awe-inspiring as sharks. The world's largest fish is a shark. One of the fastest fish is a shark. And probably the most feared fish is a shark. Some sharks really are very dangerous for swimmers and divers. Yet more people are struck by lightning every year than are attacked by sharks.

How many kinds of sharks are there? Some authorities say there are about 250 kinds, others say there are more than 350. No matter what ocean you go into, you can find some kind of shark.

Sharks are different from the bony fish, such as salmon, trout, and tuna. They have no bones. Their skeletons are made up of cartilage, the tough rubberlike material that we call *gristle* when we find it in meat that we eat.

Sharks have no swim bladders (see pages 10-13). They must swim all the time, even when asleep, or they will sink. But some sharks *do* rest on the ocean floor. They are able to do this partly because of their *spiracles.*

A spiracle is an opening some sharks have high up on each side of their heads behind the eyes. Sharks resting on the ocean floor and sucking water in through their mouths could end up with their gills clogged by sand and mud. Instead, they take water in through their spiracles, which are high enough above the bottom to keep from sucking in sand and mud along with the water.

Sharks are also known for their keen sense of smell. Scientists say that a shark can detect one drop of blood in a million drops of

water. Such sensitivity helps sharks find wounded prey, such as other fish, turtles, and seals.

Sharks also have sensitive pits on their snouts. These pits can detect the extremely weak electric currents generated by the nerves in other animals' bodies. These electric currents are less than one-millionth of a volt. In other words, the sharks home in on their prey's natural electricity.

Some kinds of sharks hatch from eggs deposited in the water. Others are born alive, as human babies are. Still others hatch from eggs inside their mothers, then

grow some more before being born. Among these last two groups of sharks, the first babies to develop inside their mother eat the others and any unfertilized eggs.

Scientists have grouped sharks into about 21 families, based on how many gills the sharks have, the shapes of their bodies, and other features.

The mackerel shark family includes the great white, mako, and basking sharks. The great white is the famous man-eating villain of books and movies—and it can be very dangerous to swimmers and divers. Although this shark is found in all the tropical and moderately warm seas, it is especially feared around Australia. The great white shark can grow to 30 feet long and weigh up to three tons.

Mako sharks, which usually avoid the shallow waters where people swim, pose less danger for humans. But makos are fierce predators that have been known to outswim and capture swordfish. Clocked at about 60 miles per hour, this shark is one of the fastest fish in the world.

The basking shark is a four-ton giant that got its name from its habit of basking in the sun at the

A great white shark (above) may grow to be 30 feet long, though most are shorter than that. These big fish eat big prey, including seals and porpoises. Usually, the great white shark makes one powerful bite, then goes away until the prey dies. Then it returns for its meal.

Shark Attack

Shark attacks on people are rare. Fewer than 100 are reported each year. But the injuries from these attacks often are very serious. This surfer on the coast of Oregon was lucky. He got away. But the shark left bites five inches deep in his back. "The entire incident lasted only about 15 seconds," the 28-year-old man recalled later.

Scientists think this attack was by a female, since female sharks are thought to be usually more aggressive than males. People have tried chemicals and other means to repel sharks. But so far, nothing works all the time. What keeps some sharks away may even attract others. Beaches where sharks have been a problem must be fenced off completely by nets offshore to be made safe for swimmers.

ocean's surface. It eats by slowly swimming with its mouth open and filtering out small creatures with its gills. A large basking shark can filter the food out of more than 1,000 tons of water in an hour.

The whale shark family has only one member. More than 50 feet long, it is the largest fish in the world (shown on pages 2-3). Like the basking shark, the whale shark filters food from the sea with its gills. Sometimes a whale shark will float with its head pointed upward just beneath the surface. As the ocean's waves pass by, the shark bobs up and down and takes in mouthfuls of food.

The smallest sharks belong to the catshark family. Most of them grow to be no bigger than three or four feet long. Some of the best-known members of this family are the swell sharks. They range through the Pacific Ocean from California and Chile to Japan and Australia. When frightened or attacked, these sharks swell up with water or air. Some inflated swell sharks float helplessly at the surface for four or five days until they can release their air and swim back to the bottom.

With more than 100 species, the requiem shark family is the largest. Most requiem sharks reach at least ten feet in length, and different ones can be found in all but the coldest seas. One kind, the bull shark, has been found far from the ocean in a river in Louisiana. Though most of these fish are harmless to people, some are known as dangerous man-eaters.

The tiger shark is the largest of the requiem family. It can grow

Its extra-wide snout (above) may improve a hammer-head shark's ability to see or smell its prey or to detect the electricity that the prey makes. A hammerhead can open its mouth a full 90 degrees so it can eat big creatures such as dolphins. The same is true of this great blue shark (right) that has just caught a mackerel.

to be more than 18 feet long, and it will attack other sharks, crocodiles in coastal rivers, and even trash thrown into the sea. Fishermen who have caught tiger sharks report finding shoes, paint cans, and license plates inside them. Because they hunt close to the shore, these sharks are considered very dangerous to swimmers.

Hammerheads are probably the oddest-shaped members of the requiem shark family. Their eyes and nostrils are located on the sides of their T-shaped snout, which some people say looks like a hammer. Scientists are not sure why the sharks' heads are shaped this way. Because their nostrils are far apart on their wide snouts, these sharks may be able to locate prey by smell better than other sharks can.

Although some of them are dangerous for people, sharks have more to fear from people than people do from sharks. After all, thousands of sharks end up on dinner menus as "marine eel" in Germany, "tuna" in France, and the "fish" of the popular "fish and chips" in England.

After growing slowly for nearly one year inside its egg case, a swell shark pup swims free. Hungry, it looks for food right away.

Animals Without Backbones

This cuttlefish, like its close relative the squid, has an internal shell instead of a backbone. This creature usually becomes active at night, when it uses its eight arms and two long tentacles to capture crabs and other prey.

Misunderstood Monsters

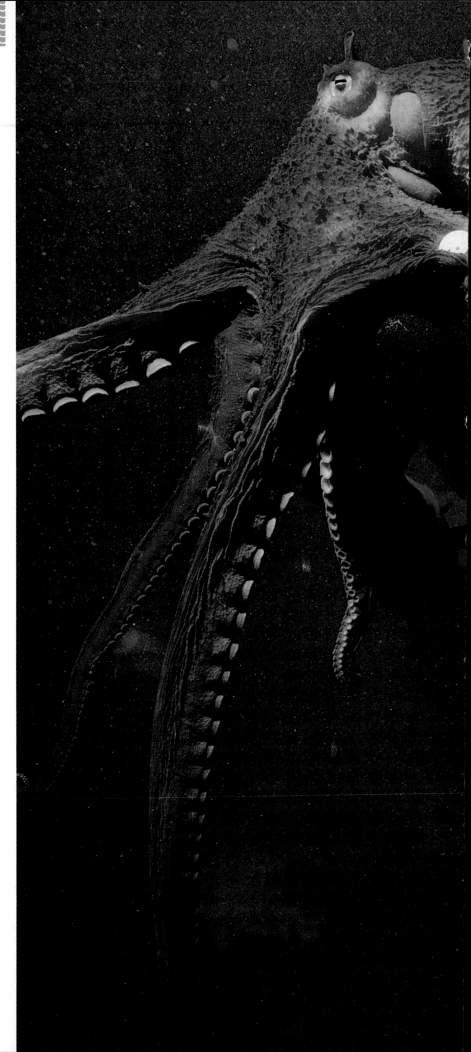

Eight long arms, lined with suckers, reached out of the ocean. One of them wrapped around a rowboat and began pull it under. But a twelve-year-old boy saved the day. He picked up a hatchet and chopped off the animal's arm. If he hadn't, the boat and its three passengers might have disappeared in the water below. What animal could have done this? It was a giant squid.

True stories like this are *very rare*. But one true story can grow into hundreds of tall tales. And, with storytellers' vivid imaginations, the "monsters" can take the shape of bigger-than-life giant squid as well as their look-alikes— octopuses, cuttlefish, and other squids.

Giant squid can grow to be 70 feet long, including over 50 feet of tentacles! They weigh several tons and have eyes the size of dinner plates. They really do look like monsters. But they feed on fish, not people. The infrequent "attacks" perhaps happen when a giant squid is startled and mistakes a boat for its enemy, the sperm whale.

Octopuses, cuttlefish, and squid are far from being monsters. Except for a few large squid and octopuses, they are rarely more than two feet long. Many are so small that they could fit on a dime. And most are really quite shy. Confronted by an enemy, they are more likely to flee than fight.

Squid, octopuses, and cuttlefish are mollusks, but they don't look at all like their mollusk cousins, the snails and bivalves. The only "shell" they make is a thin rod or sheet, called a "cuttlebone," *inside* their body. Their two large

Some of the world's largest octopuses, like the one to the left, live in the Pacific Ocean. Their bodies may be 16 inches long and their arms 16 feet long, but they are harmless to humans. Octopuses are able to change colors (below) when they are alarmed and during courtship.

eyes, which look almost like human eyes, keep a sharp lookout for prey. They can see very well, even in dimly lit water.

These "misunderstood monsters" are probably best known for their many arms. Octopuses have eight arms. Squid and cuttlefish have ten arms, two of which are twice as long as the other arms and are called tentacles. Sometimes the arms are ten times longer than the body of the animal. Rows of powerful suction cups cover the underside of the arms. Suckers on squid have sharp hooks on the inside.

Contrary to stories you may have heard, these suction cups do not suck blood. The long arms and suction cups grab and hold prey and pull it toward the beak. The beak is shaped like the beak of a parrot and is often strong and sharp enough to tear into fish and to crush lobsters. A poison in the saliva enters the prey, stunning or paralyzing it.

When an enemy gets too close, the many-armed creatures squirt out a cloud of thick, black ink and scoot off in the other direction. The ink may confuse the enemy's sense of smell. Or, because the cloud is about the size and shape of the animal that made it, the enemy may mistakenly attack the cloud and not the creature.

Have you ever seen anyone blush? Octopuses, squid, and cuttlefish "blush," too. But some can blush yellow, orange, blue, and black—as well as red. They change colors when they are alarmed and when they want to attract a mate. They also do it to hide from enemies or to ambush or even attract prey. When an octopus moves onto a pebbled area, its color becomes speckled and its skin becomes bumpy to look like rounded pebbles. A cuttlefish often lies hidden in a bed of seagrass and lures fish by making its legs white and waving them like worms.

Like a jet-powered torpedo with large fins (right), the squid darts backward to escape enemies and shoots forward to snag prey. Some larger squids cruise at speeds of 30 miles per hour. Only when it is near death (above) is the squid an easy prey for a slow-moving starfish.

Tall Tales

Because of its large size and strange shape, the giant squid has inspired many tall tales about sea monsters. Norwegians called their monster a *kraken* (CROCK-in), and Jules Verne used the same name for a tentacled creature in *Twenty Thousand Leagues Under the Sea*. In one story, the kraken was 1½ miles around. When it went back under the sea, it created a whirlpool that dragged down everything around it. Some sailors believed that the kraken could board a ship, climb into the rigging, and turn the ship over.

Living Stars

Starfish, more properly called *sea stars*, have fascinated people for many years. Oyster farmers in France once believed they were "wicked fairies" cast out of heaven. The "good fairies" were the stars that remained in the sky. In Ireland, sea stars were called the "Devil's Fingers."

What makes sea stars so interesting? Several things, from their starlike shape to the odd way they live. Though they have top and bottom sides, sea stars have no front or back, no left or right. Instead, they are built like wheels, with five to fifty arms—called *rays*—radiating from a central disc.

Sea stars travel on hundreds of tiny tube feet. The fastest sea stars have tube feet about an inch long and move about a yard per minute. Some sea stars' feet end in suction cups. The suction helps the creatures climb smooth walls.

Sea stars are carnivorous, a word that means "meat eater." They feed on a wide variety of other sea creatures, including fish, snails, and other sea stars.

Getting the meat out of a tightly shut clam or oyster shell can be a problem for people, but not for sea stars. First, they surround the shell, holding on with their strong tube feet. Then their feet start pulling. They don't have to open the shell all the way. A tiny crack no thicker than two sheets of paper is enough. The sea star then pushes its stomach in and digests the creature from the inside.

It doesn't matter if a sea star loses an arm to its enemies. It simply grows a new one—though the process, called *regeneration*, may take a year to complete. Any part of a damaged arm can be regrown. And if a broken-off arm has enough of the central disc attached, it can regrow a whole new sea star.

Regeneration proved to be a nasty surprise to some oyster farmers. The farmers wanted to get rid of the sea stars that were eating their oysters. They chopped up the sea stars and threw them back into the sea. But instead of getting rid of the sea stars, many of the chopped up pieces grew into new animals.

If an arm torn from a sea star has enough of the central disc still attached, it can grow a new body and arms (above). The underside of the arms is covered with small tubes (right). Sea stars use these *tube feet* for walking and for holding onto food.

Sea stars are found on coasts all over the world, but more kinds live along part of the Pacific coast of North America—from Puget Sound in Washington to the Aleutian Islands off Alaska—than anywhere else. Many of the more than 600 kinds of sea stars are very colorful.

Coral (below) and sponges (right) are really animals. Water enters tiny holes all over sponges. The sponges filter out bits of food floating in the water. Most kinds of coral, like anemones, have stingers that kill tiny organisms, which they eat.

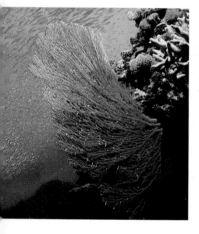

Plant or Animal?

Divers often remark on the beauty of sea anemones, sponges, and corals growing in what appear to be undersea gardens. But these are not flower gardens. Sea anemones, sponges, and corals are animals. These creatures eat, grow, and sometimes have poison that protects them.

Sea anemones do look like flowers. But a sea anemone's "petals" are wiggling, stinging tentacles. Like its relatives the jellyfish, a sea anemone uses its tentacles to capture food, including small fish, plankton, and shrimp.

A sponge has no brain, stomach, heart, or lungs. Instead, it is made up of millions of tiny cells working together. A "skeleton" of glasslike or rubbery materials gives the sponge its shape. Some cells form the sponge's outer covering. Other cells work to draw water through the many holes in the outside of the animal's body. As the water flows through, inside cells filter out tiny bits of food.

Bath sponges—the ones that are real and not plastic—come from a separate family of sponges. Such types as Dalmatian sponges, wool sponges, and others are dried in the sun. Then they are pounded and washed to remove the dead cells. Only the rubbery skeleton is left.

Sponges come in many sizes and shapes. Some weigh only a few ounces. Others may weigh more than 100 pounds. Sponges may also be round or shaped like cups, fans, plates, tubes, or even small trees.

Coral formations have many different shapes, too. They may look like branches, mushrooms, deer antlers, flat plates, and even

human brains. But whatever the shape, each formation is made up of the skeletons of animals called *coral polyps*.

Coral polyps look a lot like sea anemones. The different coral formations are created by the animals' hard skeletons. The polyps take calcium carbonate from seawater and turn it into limestone. Many use the limestone to build hard, cuplike shells around them.

When a coral polyp dies, the soft body decays. But the hard limestone skeleton remains. As new polyps build on top of the old skeletons, coral formations get larger and help to build coral reefs.

Sea anemones use hollow tentacles to capture food. Some anemones have stinging tentacles that paralyze small fish. The tentacles of other anemones are covered with mucus and, like sticky flypaper, trap small food particles floating in the water.

Homemade Homes

Over the centuries, people have admired shells for their great variety of shapes, colors, and sizes. Shells have been used for jewelry, weapons, bowls, money, and even musical instruments. But mollusks, the creatures that once *lived* in the shells, made them for a very different use. The empty shells were once their homes.

The building you live in was probably constructed by someone else, but mollusks build their own "homes." Mollusks have soft bodies and no skeleton for support or scales for protection. But they do have a fleshy cape, called a *mantle,* that surrounds much of their body. In most mollusks, this mantle produces thin layers of a chemical that hardens and slowly builds up, layer by layer, to make a shell. As long as it continues to grow, a mollusk adds on to its shell.

Snails, members of the largest group of mollusks, have shells that are built like long, spiraling tunnels. Combining shades of orange, black, brown, and white, the shells may be patterned with speckles, splotches, stripes, or zig-zags. Some may have ridges or spikes.

From the "tunnel" where it lives, the snail eases part way out to creep about at "a snail's pace" on its broad foot. If an enemy threatens, it pulls back into its shell. It often seals the opening with a toughened area on its foot.

Bivalves are the second largest group of mollusks. Their name means "two-valved," and they all have two valves, or shells, connected by a hinge. Clams, oysters, mussels, and scallops are bivalves.

The scallop (above) scoots about by flapping together the two halves of its shell. The helmet shell (left) creeps around on its broad foot. But the giant clam (far left) doesn't move around at all. It weighs more than one ton and is often wedged between patches of coral.

Shell collectors beware! The poison used by some cone shells to paralyze their prey (above) is powerful enough to kill humans.

While it is growing, a chambered nautilus (right) makes a new chamber every 35 to 40 days. A full-grown nautilus may have up to 38 chambers.

around by currents.

The chambered nautilus, an uncommon and unusual mollusk, also makes a spiraling shell. Unlike the snail, the nautilus builds shell walls to form chambers in the back part of its "tunnel" as it grows. Scientists think the nautilus slowly changes the amount of water in these chambers to help it keep the right buoyancy—so it won't sink like a stone or float at the surface like a stick of wood.

Most of the bivalves are filter feeders. Many clams, for example, suck water past their gills with a long, strawlike "neck." Their gills—which are also used for breathing—filter tiny bits of food out of the water.

Several snails eat other mollusks. Some get to their food by drilling holes in their prey's shell. Some soften the shell with acid. They then pull the meat out with their tongue, which is covered with tiny, sharp teeth.

Some snails, called *cone shells*, thrust their tongue out like a spear to capture worms or small fish. A poisonous tooth shoots out from the tongue, paralyzing the prey instantly.

The nautilus has a sharp beak, like a parrot's beak, to tear apart fish and crush crabs and lobsters. It seems to be really fond of eating old exoskeletons left behind by molting lobsters.

For protection, a bivalve closes its shell and holds it tightly shut with two very strong muscles.

Scallops, and a few other bivalves, flap the halves of their shell to swim by a kind of jet propulsion. Most bivalves, however, stay put. Mussels anchor themselves with many threads to something firm, like rock. Oysters cement themselves to rocks and other oysters. Many clams, which often live buried in sand or mud, dig their foot downward as an anchor. The giant clam has no anchor. Reaching a length of nearly four feet and weighing over 2,000 pounds, it is too heavy to be moved

Pearl Farming

A natural pearl is created when a grain of sand or other small object gets inside the shell of a pearl oyster or other mollusk. The mollusk coats the tiny invader with layers of minerals. A natural pearl, which may grow for 10 to 12 years, is pearl all the way through. But today, most pearls are grown on underwater farms. Pearl farmers place beads carved from shells inside the oysters. In two or three years the oysters coat the beads with thin layers of pearl. These pearls are called *cultured* pearls and they are not as valuable as natural pearls.

Rainbows of the Sea

People expect to see fancy fish, colorful corals, and bright anemones in the sea. So it may be a big surprise to learn that some of the most beautiful ocean animals are worms and slugs. In case you are wondering, these creatures *are* relatives of the worms and slugs on land.

Fanworms are named for the feathery fans they spread out to gather food in the water. The worms themselves live inside tubes that they build on rocks and corals. When the fans are spread, hairlike projections on the fans sweep tiny plants and animals toward the worms' mouths. But when large creatures come by, the fans snap back into the safety of the tubes.

Sea slugs, like their relatives on land, have no shells. But these colorful creatures are not as gentle as they look. They are active predators. Some devour sponges, corals, and even poisonous anemones.

The bright colors of sea slugs warn other predators that the slugs taste bad. But if enemies try to attack them anyway, some slugs have a really tricky defense. Their diet includes anemones and jellyfish, creatures with stinging cells. These cells pass harmlessly through the slugs' digestive systems. Then the cells are stored in the tips of the soft "horns" covering the animals' bodies. Any animal that attacks the slugs gets a mouthful of stingers.

These strange "horns" are really gills. The slugs breathe through them. Because the gills are outside their bodies, sea slugs are also called *nudibranchs* (NOO-di-branks), a name that means "naked gills."

Christmas tree worms live in tubes they bore into coral reefs. The flowery tentacles (above) serve as gills and collect food. Waste is carried out through a special groove in the tube. When a fish comes too near, the worm pulls its tentacles into the tube (right).

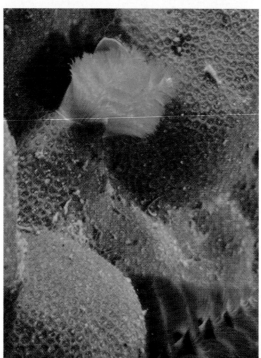

Many sea slugs glide on a large foot (left, bottom), but on some sea slugs the foot has flaps (left, top) that are used for swimming. The stalks on their heads contain eyes and are sensitive to smells and touch. Sea slugs lay their eggs in ribbons (below) and clumps.

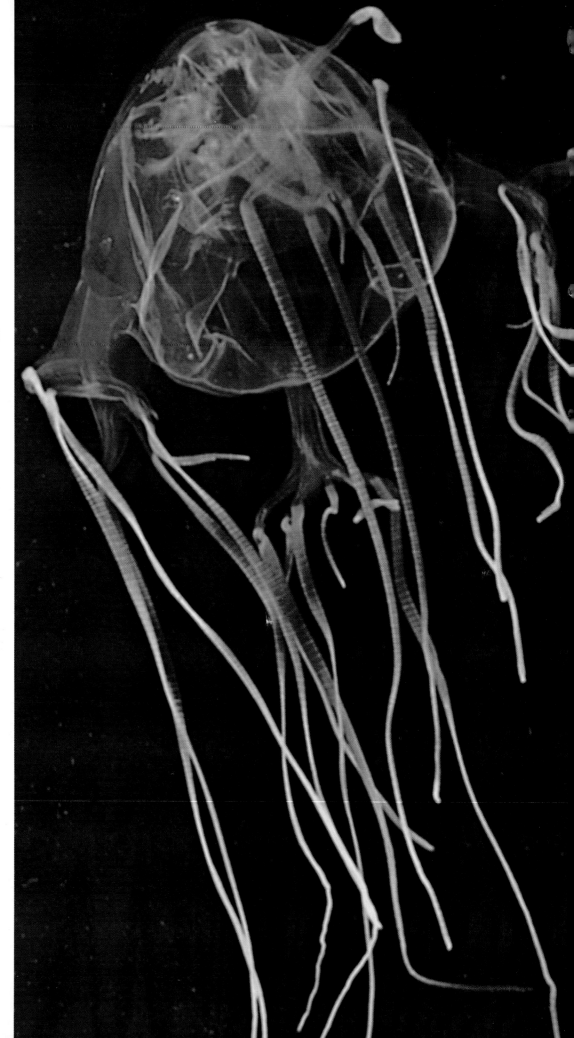

The sea wasp jelly-fish (right) stings far worse than any wasp does. The sea wasp is common off the coast of Australia. More people have died from stings from its tentacles than from attacks by sharks.

The Drifters

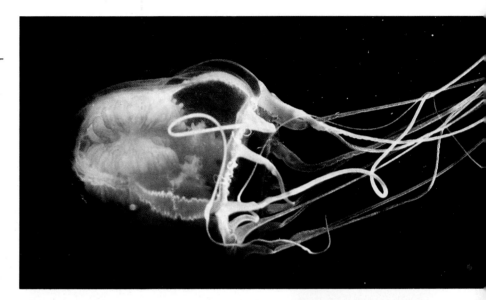

Jellyfish and their look-alike distant relatives called *hydrozoans* aren't made of jelly—and they aren't fish, either. They have no hearts, no brains, no bones, and no fins. In fact, these creatures are almost all water. Even so, some of them deliver a sting more powerful than that of a wasp.

Most jellyfish and hydrozoans come in two main forms. The form called a *polyp* (PAHL-ip) looks like a tiny anemone. Polyps anchor themselves to anything solid —rocks, piers, even rusty cans. The familiar bell-shaped jellyfish is the form called a *medusa*. It is named after the Greek goddess Medusa, whose head was covered with snakes instead of hair.

Some jellyfish and hydrozoans drift with the tides. Others swim with a form of jet propulsion. They squeeze their bodies to force water out of the open part of their bells. The flowing water pushes the creatures along very slowly.

The Portuguese man-of-war, a hydrozoan, moves in a different way. It has a bright blue float that stays on the surface and serves as a sail, moving along easily with the slightest breeze.

Whether they swim or sail, jellyfish and hydrozoans are called *drifters* because they move with the water currents.

Traveling so slowly creates no problems for these animals. They move to keep their tentacles untangled and spread out to catch food—fish, shrimp, young crabs, and even other kinds of jellyfish.

When prey brushes by, it is struck by tiny threads that pop out from special cells in the tentacles.

Some threads are like hollow needles, injecting poison to paralyze the prey. Others are like ropes that hold the prey in place.

The most dangerous tentacles are those of the sea wasp, found off the shores of Australia. These tentacles carry some of the most powerful venom in nature. A sea wasp's sting causes painful blisters and can be fatal. People who survive the sting may be left with scars.

Although jellyfish can be harmful to people, some are helpful, too. People in the Far East use them as food, either deep-fried or steamed. Doctors use chemicals from jellyfish to treat illnesses, including jellyfish poisoning.

Both these creatures are jellyfish, but the one on the bottom doesn't have long tentacles for capturing prey. Instead, it has many cauliflower-shaped arms which act as mouths. These "mouth-arms" trap bits of food in the water and pass them into the jellyfish's stomach.

Knights
in Armor

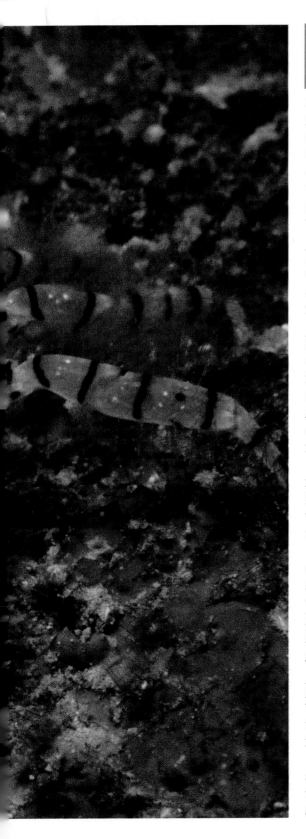

Like knights of old, some sea creatures are protected by hard suits of armor. Only this armor is natural, and as the animals get bigger they shed it and grow more. Among the best known armored creatures are crabs, lobsters, shrimp, and barnacles. Their armor is like a shell, and we call these animals *crustaceans* (kruh-STAY-shuns), from a Latin word that means "shell."

A crustacean's shell serves as both its skin and its skeleton. Scientists call this kind of shell an *exoskeleton*, from "exo," which means "outer."

The exoskeleton doesn't grow, but the animal inside does. Whenever the crustacean gets "too big for its britches," it splits its shell and struggles out. This is called *molting*. The crustacean usually eats the old exoskeleton, which is a rich source of food.

Crabs can be found in the north polar region and in the tropics, on land, in trees, in shallow water, and in the deepest parts of the ocean.

Most crabs live in the ocean, crawling on the bottom. They usually walk sideways, some on only two legs, some on all ten. Some, like the blue crab, have one pair of legs shaped like paddles and are good swimmers.

The largest crab is the Japanese spider crab. It lives in deep waters off the coast of Japan. From tip to tip of its longest legs, it may measure up to 26½ feet.

Most crabs have eyes on stalks. The stalks can bend and turn, allowing the crab to look in almost any direction. In moments

The pom pom crab (left) holds up two stinging anemones to ward off any hungry attackers. The coral crab (left, above) relies on its extra-thick shell to save it from all but the strongest jaws and the sharpest teeth. A crab that has just molted (above) is soft and sluggish. For it, the best protection from an enemy is a good hiding place.

of danger, many crabs burrow in the sand or mud. Some others cover their backs with sponges or anemones, trying to avoid danger through camouflage.

Have you ever seen lobsters in a tank at a supermarket? They were probably American lobsters, five or six years old, and weighed about one pound. The American lobster can live up to 100 years. At that age it might weigh as much as 50 pounds and be more than two feet long.

Except for their color, shrimp often look like little lobsters without claws. They come in almost every color of the rainbow. Some are almost transparent, like glass. Look closely at a "glass shrimp" and you can see its heart beat or watch its food go down.

One shrimp is called a *weaver shrimp* because it weaves threads of seaweed together to make a tube-shaped home. A pistol shrimp has one claw that is almost as large as the rest of its body. Part of the claw operates like the firing hammer of a pistol, making a popping noise. The noise probably warns other pistol shrimp to stay away.

A barnacle looks more like a little fortress than a knight in armor. The shell we see is not an exoskeleton but a thick covering made of the same material as a clam shell. Snug inside this fortress is what one person called a "little shrimplike animal standing on its head." All you may ever see of the animal itself are its twelve feathery legs fluttering in the water to filter out small bits of food.

Some barnacles grow no larger than your little toe. But one kind, found off the western coast of South America, can grow to be as large as a pineapple. No matter what their shape or size, all barnacles are attached firmly and permanently to something—rocks along the shore, the bottom of a boat, or even the head of a whale.

Where's our dinner? Cemented permanently to one spot, goose barnacles (top) count on ocean currents to carry tiny food particles their way. Lobsters, like the spotted reef lobster (middle) and the spiny lobster (bottom), must search for their next meal— a clam, a starfish, or even another lobster.

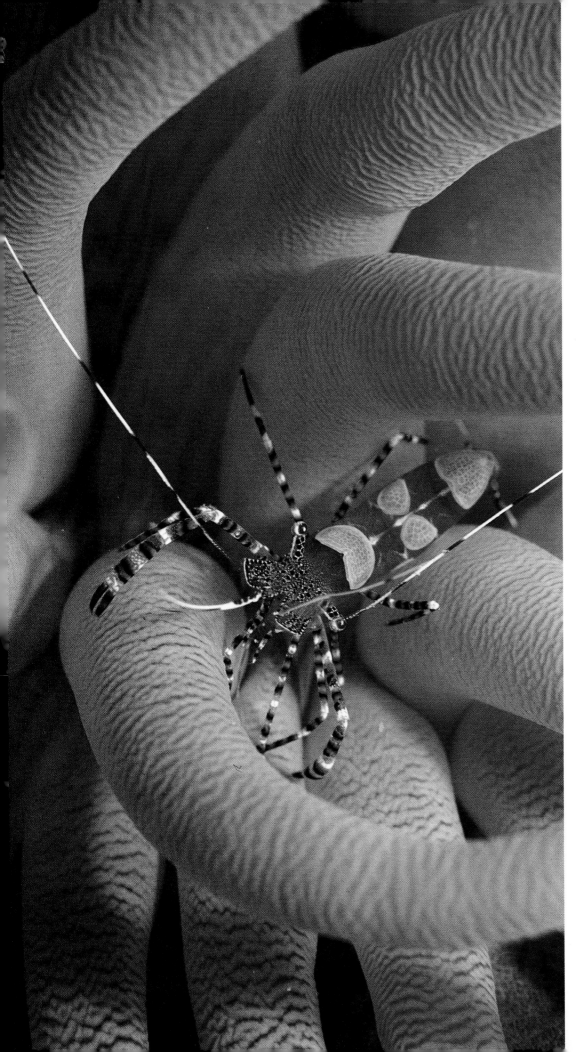

The spotted cleaner shrimp on an anemone (left) and the banded shrimp (below) pick parasites from the scales, fins, and gills of some fish. These shrimp provide a service and get a meal at the same time.

Sea Mammals and Reptiles

Sleek swimmers like these Atlantic spotted dolphins appear perfectly suited to underwater life. But like all sea mammals or reptiles, they must surface to breathe.

Performers and Predators

Dolphins must be trained to jump through hoops and perform other acrobatics in marine shows (right). In the wild, dolphins often swim and leap in front of boats as if they were racing or leading the way (above).

Dolphins fascinate people. Crowds at marine shows cheer as these sleek creatures twist and turn and leap through hoops. The true story of a wild dolphin leading ships safely through a narrow channel has been printed in many books and magazines. So have stories about other dolphins swimming close to shore and letting children pet them and ride them. Scientists have even tried to teach dolphins to talk, but without success. What are these creatures that seem so playful, so friendly, and so intelligent?

Dolphins are mammals, just as cats, cows, and people are mammals. Like other mammals, dolphins must breathe air—even though they live in water. A dolphin swims to the surface to breathe out and in through its blowhole, a small hole in the top of its head.

Dolphins sleep for only five or six minutes at a time, sometimes with one eye open. If they were to sleep too long they would sink and drown. If one dolphin is sick or injured and starts to sink, other dolphins will support it at the surface for hours or even days.

More than 30 kinds of dolphins inhabit the world's oceans from the Arctic to the Antarctic. Bottlenosed dolphins are probably the best known. Named for their bottlelike snouts, they are the star attraction in many marine shows. Flipper, the dolphin seen in a television series, was a bottlenosed dolphin. Spinner dolphins, which are common around Hawaii, can be spotted performing their leaps and somersaults in the wild. Other

Killer whales, also known as *orcas*, eat other dolphins, seals, fish, and even whales. Sometimes the orcas chase their prey right out of the water, as seen here.

dolphins must be trained to perform such feats.

One member of the dolphin family is orca, the "killer whale." Orcas may grow to be 32 feet long, making them the largest of all dolphins. It's easy to spot a male killer whale; when it surfaces to breathe, its six-foot-high dorsal fin juts out of the water.

Most dolphins have more than 100 teeth, which they use for holding their prey—but not for chewing. When a dolphin goes fishing for a meal of small fish, squid, or shrimp, it grabs its food firmly. Then, with a gulp, it swallows the creature whole.

All dolphins have very large brains, even larger than most human brains. Since the animals also learn quickly and some even seem to like being around humans, people have suggested that dolphins are as intelligent as humans. But many researchers now believe that the dolphins' large brain size is related to their keen hearing.

Although their ears are only tiny pinholes, dolphins can hear very well. They can hear pitches far higher than those humans can detect. Dolphins often search for food and find their way by making high-pitched clicking sounds which bounce off other objects and create

echoes. A dolphin's large and sensitive brain interprets these echoes to reveal the size, shape, and speed of the objects around it.

Some people use the names "dolphin" and "porpoise" to refer to the same creatures. Porpoises do look a lot like dolphins, but some scientists separate the two groups into separate families. Porpoises lack the "beak" common to most dolphins. A porpoise's teeth are usually flattened at the tips. A dolphin's teeth are pointed. Porpoises usually travel in small groups of only two to four, but up to 2,000 dolphins may gather in one area where the fishing is good.

Splashing out of the water is called *breaching,* and orcas do it often. Breaching may be a signal to other orcas, and the heavy splash may also stun schools of fish that the orcas are chasing. Female orcas, like the mother at left with her calf, nurse their young for about a year. Females sometimes wait up to ten years between having calves.

To get this close-up picture of a gray whale's eye, the photographer had to swim close enough to touch the giant animal. The whale's eye is protected against the irritation of salty sea water by secretions from tear glands.

The Giants

The largest animal that ever existed is alive today—and it lives in the sea. Its tongue weighs as much as an elephant and its heart is the size of a small car. Its young grow about 200 pounds a day. Eventually, it grows to be around 90 feet long—longer than five large cars sitting bumper-to-bumper. What is this giant? It is the blue whale, the largest of all whales.

Whales seem to have evolved from animals that moved from the land back into the sea. Whales are well-adapted to their life in the ocean. Their nostrils, which once were located on the front of their noses, have shifted to the top of their heads. These nostrils are called *blowholes.* Blowholes let the whales breathe as they skim the ocean's surface. Under water, the blowholes shut tight so the whales don't drown.

Whales' shapes also show how these animals are adapted to life in the sea. Their front legs have turned into flippers, their hind legs have disappeared, and their tails have developed wide flukes. These changes help the whales swim better. Their tails propel them through the water; their flippers help them change direction; and their streamlined shapes cut down water resistance, making swimming easier.

Unlike most land mammals, whales do not have hair or fur to keep them warm. Long fur would make swimming more difficult. Instead, whales have developed a layer of fat, called *blubber,* which helps keep out the cold. Bowhead whales, which live in the Arctic, have blubber almost two feet thick.

Mother humpbacks use strong fins to help them protect their babies, called *calves*. At birth, a calf weighs about two tons. It stays with its mother for nearly a year.

Scientists divide whales into two groups: baleen whales and toothed whales. Instead of teeth, baleen whales have long filters hanging from the insides of their jaws. These filters, called *baleen*, look like giant combs.

Baleen whales are the biggest whales, yet a large part of their diet is plankton, which includes the tiniest creatures in the sea. Some baleen whales swim with their mouths open through swarms of plankton, letting the baleen filter out their food. Other baleen whales take in mouthfuls of water filled with plankton. When they close their mouths, the water strains out through the baleen and leaves the

food behind. A large whale may take in four tons of food a day.

The largest baleen whale is the blue whale. The second largest is the fin whale, about 80 feet long. That's longer than a tennis court. The smallest is the pygmy right whale. It usually measures only about 15 feet long and could fit inside many living rooms.

Toothed whales are smaller than most baleen whales, but there are many more kinds of them. Dolphins and porpoises are toothed whales. Sperm whales, which may be more than 65 feet long, are the largest of the toothed whales.

Sperm whales dive deeper than any other whale. One sperm whale

Strands of baleen filter plankton from the water as a humpback whale feeds. Humpbacks feast during the summer. But they go without food in the winter.

was found tangled in an underwater cable more than 3,700 feet down. Sperm whales eat deep sea squid, so they must search deep water for their food. When a sperm whale tangles with a giant squid, the squid gets eaten. But the whale, though larger than the squid, does not escape injury. Its head is often scarred by suction cups on the squid's long tentacles.

Sometimes when a whale eats a squid, the squid's hard beak lodges in the whale's stomach. A thick, greasy substance forms around the beak. Later, the whale may get rid of this material, which is called *ambergris.* Lumps of ambergris found in dead whales or

A beluga stays under water only about five minutes before surfacing to breathe. The large bulge on its head may help direct the whale's "sonar."

floating in the sea are used to make expensive perfume.

Narwhals (NARR-walls) have the largest teeth of all toothed whales. One of a male narwhal's upper teeth grows straight out like a horn, sometimes measuring eight or nine feet long. Early whalers claimed that these were horns of the mythical unicorn and had magic powers. Even today people call narwhals the "unicorns of the sea."

Like dolphins, most of the other toothed whales seem to use high-pitched sounds to find their way and to locate fish and other prey. Belugas, or white whales, sound so much like birds they have been called "sea canaries." But many of the belugas' sounds are too high-pitched for humans to hear.

Finding food and other objects by sound is called *echolocation*. The animals creating the sounds listen to the echoes bouncing off objects. Echoes reveal where the prey is and how fast it is moving.

Whales have few enemies. Only killer whales, largest members of the dolphin family, seem to prey on whales regularly. Because they hunt in groups, sometimes up to 40 strong, killer whales are able to overcome whales that are much bigger than they are.

A whale's greatest threat is a human whaler. Over the centuries, whalers have drastically reduced the numbers of many kinds of whales. Today, blue whales are endangered—in danger of becoming extinct—although their numbers do seem to be increasing in a few places. Right whales are endangered, too. These whales were named by whalers, who said the animals were slow and easy to kill —the "right" whales to hunt.

Today, it is illegal to kill many types of whales, and most countries have abandoned whaling altogether. But other nations continue the hunt, taking the animals for their meat and for their oil.

Whale Song

Probably the most famous sound of the sea is the song of the male humpback whale. These whales may sing for hours, repeating the same "tune" over and over. The sounds, which have been compared to snores and screams, can be detected with scientific equipment hundreds of miles away. People have many questions about the songs. Are they only a signal from one male to another, or do they help the males attract mates? Is a whale born knowing just one tune, or can he choose what he sings?

The Real Mermaids?

Several years ago, a ship's captain sighted what he thought was an accident at sea. He saw three people in the distance bobbing in the water and about to drown. He signaled the group and approached it. But when he got near, the three slid under the water and swam away. The captain realized that he had not seen people at all. Instead, he had spotted a group of large marine mammals called *dugongs.*

From a distance, dugongs and their relatives, the manatees, seem to resemble humans swimming. Could these animals be the mermaids of legend? Mermaids were sea creatures that were supposed to be half human and half fish. No one is sure what people who said they saw mermaids *really* saw. But some seafarers who reported seeing mermaids may have seen dugongs.

Up close, dugongs and manatees don't look like people at all. To start with, they are much larger. Some measure up to 13 feet long and weigh more than a ton. Both have thick, wrinkled skin, and manatees' flippers have nails that look like an elephant's toenails. In fact, elephants are these animals' closest relatives on land.

Dugongs and manatees spend their entire lives in water, either at sea or in rivers near the ocean. Dugongs stay in the coastal waters of eastern Africa, Asia, Australia, and New Guinea. Manatees travel along the coasts and in coastal rivers of western Africa, North and South America, and a few islands in the Caribbean Sea.

Both creatures eat only plants, pushing underwater grasses into

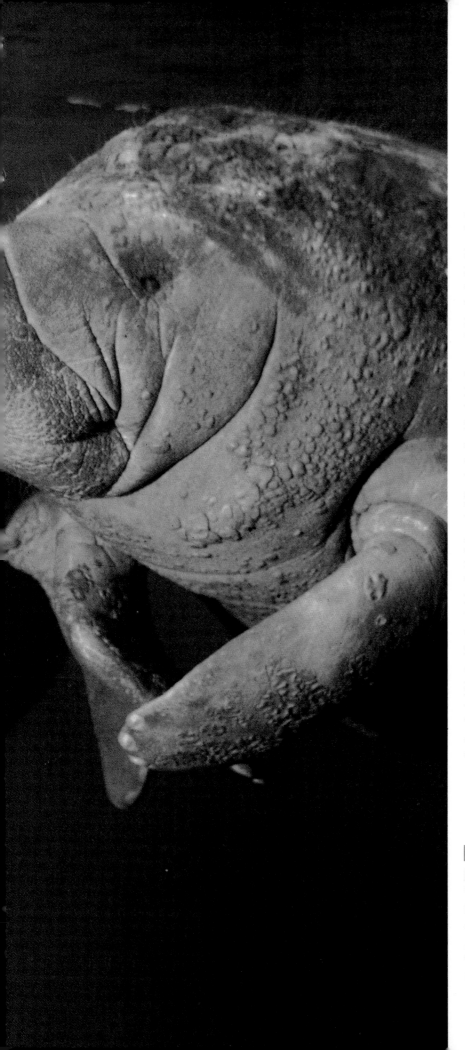

their mouths with their giant, bristle-covered lips. When not eating, the large, slow animals spend most of their time resting. They drop to the bottom of shallow waters or float just beneath the surface. Every few minutes, they bob up to breathe.

Because of their large size, dugongs and manatees have few natural enemies. Yet they are in trouble almost everywhere. In Florida, manatees are no longer hunted and are protected by law. But, because they often swim just below the surface where boaters cannot spot them, many are injured and killed by power boat propellers.

In the southwest Pacific and Indian oceans, dugongs are hunted frequently for their meat, oil, and tusks. Though many animals remain, their numbers cannot hold up under continued hunting. They could meet the same end as their relative in the far north, Steller's sea cow. That giant was nearly 25 feet long, as large as a small whale. Its thick skin looked like the bark of a tree and was as tough as an automobile tire. It was hunted to extinction only about 200 years ago.

Rubbing noses is one way manatees communicate with each other. But these large mammals usually keep to themselves, swimming slowly and eating plants.

Creatures of Sea and Shore

Seals of different kinds live along many coasts throughout the world. But few face harder living conditions than those in the north and south polar regions. Walruses, close relatives of the seals, live *only* in the Arctic Ocean—where the water is cold.

The Weddell seal's home is Antarctica, where temperatures in the winter sometimes drop to 76° below zero F at sea level. Weddell seals escape this bitter cold by staying in the water under solid ice all winter. The sea water temperature there is around 29° F—and that's just right for these creatures.

Seals and walruses don't get cold in nearly freezing water. Why not? One reason is that they are protected by a thick layer of fat and oil called *blubber* that holds in body heat.

Another reason that seals and walruses keep warm is the shape of their bodies. A human being's arms and legs add a lot of surface area to his body, and that allows lots of body heat to escape. But a seal's or walrus's flippers only make up a small portion of the area of its body. Little heat escapes through the flippers.

Some kinds of seals and all walruses control the amount of blood that flows through their thick flippers. Blood is warm. When it flows close to the surface of skin it loses warmth if the air or water next to the skin is cooler than it is. By reducing the flow of blood to the surface, these seals and walruses keep their blood from cooling.

Fur also helps seals keep warm, especially fur seals, which have more hair than other seals do.

All these seals are called "eared" seals because little flaps cover their ears. Eared seals, like sea lions (opposite and left) and fur seals, come to beaches to breed. Whenever their mothers go into the ocean to eat, fur seal pups (above) form groups and wait for them to return.

Both male and female walruses have tusks. Tusks are used for many purposes, including chopping ice, helping climb onto ice, and fighting. In groups of walruses, the walruses with the biggest tusks are the most important. The tusks of older walruses often are broken or worn down.

The hair is made oily by special glands in the skin. The oily hair repels water and traps an insulating layer of air next to the body.

Seals and walruses are big mammals. Male elephant seals, the largest of them all, can weigh over 4 tons. They need a lot to eat!

Walruses mostly eat clams and other mollusks. Many seals eat whatever seafood is easiest to catch. But some seals are very picky eaters. Crabeater seals eat almost nothing but Antarctic krill, shrimplike crustaceans. Southern elephant seals feed mostly on squid. And leopard seals catch lots of penguins and other seals in the Antarctic waters where they live.

Seals and walruses dive to find food. When they go under water, they close their nostrils and their ears. Some kinds of seals seldom stay under water more than five minutes, but other kinds can stay under for over an hour.

Walruses push their snouts along the ocean floor to find food. Their snouts have about 450 sensitive whiskers that help them know when they have located something to eat.

Walruses' big tusks are not used to locate food, as people once thought. They are useful in making their owners important. The walrus with the biggest tusks gets the best places to rest on the shore or ice and wins the right to mate with more females.

Most seals and walruses stay in the water most of the year, but the females go onto ice or land to give birth. They usually have only one pup, which grows very fast. During breeding season, more than

A mother sea otter
has caught a sea
urchin for her hungry
pup waiting for her to
return to the surface.

a million fur seals may haul out onto a single group of islands.

Sea otters live in the Pacific Ocean and seldom come to shore. They even give birth in the water. Yet, sea otters usually don't go more than about a mile away from the shore.

Sea otters are smaller than most seals and other marine mammals and do not have any blubber. They are about three feet long, not including the tail. Because they are small, they lose body heat more quickly than the larger marine mammals do. That is why sea otters' very dense fur is important for keeping warm. Small mammals also burn food faster than large mammals do. Burning food makes

energy and this energy helps sea otters stay warm. But it also means they have to eat a lot.

Adult sea otters eat 15 to 20 pounds of food a day, almost a quarter of their own weight. They gather food by diving to the shallow bottom along the coast. Their favorite meals are sea urchins and mollusks such as abalones.

Once a sea otter has found food, it usually takes it to the surface to eat it. Sometimes otters carry the food in their front paws. Other times they stuff it into a pouch of loose skin at the armpit. Often, they also put a stone in the pouch. When an otter reaches the surface, it flips onto its back and begins its meal. Food it can't crunch between its teeth is broken open by hitting it on the stone.

Many, many fur coats have been made out of sea otter fur. By the early part of this century, nearly all the sea otters had been harvested to get their fur. In 1911, several countries agreed to regulate the hunting of seals and otters, even though no one knew if any sea otters were left to protect.

Miraculously, a few sea otters did survive off the icy shores of Alaska and the rocky beaches of California. Today, there are again thousands of sea otters off these two coasts.

This young sea otter resting on its mother's chest is waiting to be groomed. It won't start diving until it is about two months old. **Although it may eat solid food soon after it is born, it also takes milk from its mother, maybe until it reaches adult size in six to eight months.**

Slither, Swim, and Crawl

Crawling out of the shell, a baby green sea turtle prepares for its scramble across the sand and into the sea. Adult green sea turtles (right) spend their lives in the water except for a short period about every three years when the females return to beaches to lay eggs.

No book about sea creatures would be complete without mentioning some of the seafaring reptiles: the sea turtles and sea snakes. Sea turtles spend most of their time in the oceans. But every few years, the turtles gather and return by the thousands to the shore to lay their eggs. Some return to the same beaches each time, possibly the very beaches where they themselves were hatched.

How do the turtles find each other? How do they find the same beach, year after year? No one knows for sure. The turtles may recognize and follow familiar smells in the ocean currents.

Some sea turtles grow to enormous sizes. The leatherback, which has a hard, leathery skin instead of a shell, is the largest. It grows to be six or seven feet long, as long as your bed might be.

Sea turtles can also live a long time, as much as 100 years or so. Unfortunately, not many turtles live to be 100 anymore. Some of the turtles are hunted for their meat, their hide, their shells, and for a substance inside the shell called *calipee*, used to make turtle soup. For this reason, one expert has called the green sea turtle, the favorite among chefs, "the most valuable reptile in the world."

Gulls, vultures, and other animals also threaten the turtles. Raccoons find the eggs a special treat. And seabirds grab the newly hatched youngsters as they lurch out of their eggs and struggle across the sand to the sea.

Divers report that some sea animals are as curious about people as people are about the animals. Problems come up when a nosy creature wrapped around a diver's arm or leg turns out to be an inquisitive, but poisonous, sea snake.

Some sea snakes are so slow to bite that divers can handle them without being bitten. Others are more aggressive. But even when they bite, the deadly snakes may not use their venom. They need the venom to capture fish and other prey for food. A hard but nonpoisonous bite is their way of saying, "Leave me alone!"

Sea snakes stay under water longer than most other animals that breathe air, often two hours or more. They are helped by their extra-long lungs, which run three-fourths the length of their bodies. A small sac at the tip of the lung stores even more air. And the snakes can reduce their need for air by cutting their pulse rates in half.

Two types of sea snake range through parts of the Pacific and Indian oceans. The Atlantic Ocean has no sea snakes at all. The snakes known as *sea kraits* stay around reefs and coastlines. They come ashore to find food and to lay eggs. The other snakes, called *true sea snakes*, bear live young and never leave the sea.

Sea snakes have many different diets. Some survive on fish eggs. Others dive and attack eels and small fish on the ocean floor. The yellow-bellied sea snake floats at the surface with its head and tail hanging down like bait on a hook. When a curious fish comes near, the snake strikes.

Sea snakes seem to have only two enemies: sea eagles and people. Hungry eagles catch the snakes on the surface and drop them on nearby rocks to kill them. And in some places, people cook the snakes and sell the skins.

Most fish leave sea snakes alone. Even sharks are said to avoid these reptiles. Any fish that tries to catch and eat a sea snake stands a good chance of being bitten.

With one of the deadliest bites on earth, poisonous sea snakes are well able to protect themselves from undersea predators. Some kinds of sea snakes spend their entire lives at sea and are said to be able to hold their breath under water up to eight hours.

Index

Illustrations are in **boldface** type.

Credits

Cover: (clownfish and anemone) Howard Hall. **Half title page:** (angelfish) Alex Kerstitch. **Title page:** (whale shark and diver) Howard Hall. **Contents:** (octopus) Alex Kerstitch. **Page 6:** Howard Hall. **7:** Doug Perrine/DRK Photo.

Fascinating Fish

8-9: Dwight Smith/Mandojana & Assoc. **10:** Doug Perrine/DRK Photo. **11:** Howard Hall. **12:** Jeff Foott. **13:** clockwise from top left: Jeff Foott, Howard Hall, Carl Roessler, Zig Leszczynski/Animals Animals. **14-15:** left, Zig Leszczynski/Breck P. Kent; right, Rudie H. Kuiter. **16:** left, Michael DiSpezio; right, Virginia P. Weinland/National Audubon Society Collection/Photo Researchers. **17:** G. I. Bernard/Oxford Scientific Films. **18-19:** Howard Hall. **20:** Carl Roessler/Tom Stack & Assoc. **21:** David Doubilet. **22:** top, Carl Roessler; bottom, Douglas Faulkner/Sally Faulkner Collection. **23:** top, Jane Burton/Bruce Coleman Inc.; middle, Zig Leszczynski/Breck P. Kent; bottom, Douglas Faulkner/Sally Faulkner Collection. **24:** Warren Williams/Seaphot. **25:** clockwise from top: Alex Kerstitch, Ed Robinson/Tom Stack & Assoc., Zig Leszczynski/Breck P. Kent, Zig Leszczynski/Breck P. Kent, Alex Kerstitch. **26:** top, Doug Wechsler; bottom, Runk/Schoenberger/Grant Heilman. **27:** Ed Robinson/Tom Stack & Assoc. **28:** Howard Hall. **29:** Christopher Newbert/from the book *Within A Rainbowed Sea*, Beyond Words Publishing Company, Rte. 3, Box 492B, Hillsboro, Oregon 97123. **30:** Carl Roessler. **31:** J. H. Carmichael/Bruce Coleman Inc. **32:** left, Christopher Newbert; right, Fred Bavendam. **33:** Howard Hall. **34:** Georgette Douwma/Seaphot. **35:** left, Georgette Douwma/Seaphot; right, Doug Perrine. **36-37:** all by Peter David/Seaphot. **38-39:** Alex Kerstitch. **40-41:** Timothy O'Keefe/Bruce Coleman Inc. **41:** left, Dr. E. R. Degginger; right, Christian Petron/Seaphot. **42:** Jeff Rotman. **42-43:** Howard Hall. **44:** Sam Foster. **45:** both by Howard Hall. **46-47:** Koji Nakamura.

Animals Without Backbones

48-49: Carl Roessler/Tom Stack & Assoc. **50-51:** Fred Bavendam. **51:** both by Jeff Rotman. **52:** Ron Church/Tom Stack & Assoc. **52-53:** Peter David/Seaphot. **53:** Bettmann Archive Inc. **54:** top, Jeff Rotman; bottom, Gary Milburn/Tom Stack & Assoc. **55:** clockwise from top left, Heather Angel, Zig Leszczynski/Animals Animals, Douglas Faulkner/Sally Faulkner Collection, Douglas Faulkner/Sally Faulkner Collection, Dr. E. R. Degginger, Keith Gillett. **56:** left, Howard Hall; right, Robert Holland/DRK Photo. **57:** Howard Hall. **58:** Howard Hall. **59:** top, W. H. Amos/Bruce Coleman Inc.; bottom, Chris Prior/Seaphot. **60-61:** clockwise from top left: Alex Kerstitch, Howard Hall, Breck P. Kent, Douglas Faulkner/Sally Faulkner Collection, William A. Conklin. **62:** both by Howard Hall. **63:** clockwise from top: Neil G. McDaniel/Tom Stack & Assoc., Brian Parker/Tom Stack & Assoc., Alex Kerstitch. **64:** Keith Gillett/Tom Stack & Assoc.

65: top, Larry Madin/Seaphot; bottom, Christian Petron/Seaphot. **66:** top, Fred Bavendam. **66-67:** Fred Bavendam. **67:** right, David Maitland/Seaphot. **68:** all by Alex Kerstitch. **69:** left, Fred Bavendam; right, James H. Carmichael, Jr.

Sea Mammals and Reptiles

70-71: Howard Hall. **72-73:** left, Howard Hall; right, Robert L. Dunne. **74-75:** Jeff Foott/Tom Stack & Assoc. **75:** both by Jeff Foott. **76-77:** Howard Hall. **78-79:** Flip Nicklin. **79:** bottom, Charles and Peter Jurasz. **80:** Zig Leszczynski/Animals Animals. **81:** top, Fred Bruemmer; bottom, Mark J. Ferrari, Center for Whale Studies. **82-83:** Jeff Foott. **84:** Fred Bruemmer. **85:** top, Stephen J. Krasemann/DRK Photo; bottom, Howard Hall. **86:** Bryan Alexander/Black Star. **87:** George Holton. **88-89:** both by Jeff Foott. **90-91:** left, David Hughes/Bruce Coleman, Inc.; right, Island Resources Foundation. **92-93:** Howard Hall.

Library of Congress Cataloging-in-Publication Data

Ranger Rick presents Amazing creatures of the sea.

Includes index.
Summary: Describes the physical characteristics, habits, behavior, and interrelationship of a variety of fish, invertebrates, mammals, and reptiles that live in water.
1. Marine fauna—Juvenile literature. [1. Marine animals] I. National Wildlife Federation. II. Title: Amazing creatures of the sea.

QL122.2.R36 1987
591.5'2636 87-20413

ISBN 0-912186-83-6

Acknowledgements

Creating this book involved many people with an interest in underwater life. We wish to thank the photographers whose work appears in this book; their first-hand accounts of marine life were an invaluable resource. We are also grateful to the staff members of the Smithsonian Institution's National Museum of Natural History, who willingly answered our questions. And, in particular, we wish to thank Dr. G. Carleton Ray, Research Professor, Environmental Sciences Department, University of Virginia, whose many comments and suggestions were extremely helpful.

National Wildlife Federation

Jay D. Hair
President

James D. Davis
Senior Vice President for Membership and Support Programs

Alric H. Clay
Senior Vice President for Administration

William H. Howard Jr.
Senior Vice President for Conservation Programs

James R. De Santis
Vice President for Public Affairs

Francis A. DiCicco
Vice President for Financial Affairs

Lynn A. Greenwalt
Vice President for Resources Conservation

S. Douglas Miller
Vice President for Research and Education

Joel T. Thomas
General Counsel

Staff for this Book

Howard F. Robinson
Editorial Director

Victor H. Waldrop
Project Editor and Writer

Debby Anker
Illustrations Editor and Writer

Elizabeth B. Blizard
Betty Olivolo
Researchers and Writers

Donna Miller
Design Director

Tim Kenney
Designer

Michele Morris
Editorial Assistant

Paul Wirth
Quality Control

Margaret E. Wolf
Permissions Editor

Tina Isom
Holly Ritland
Production Artists